BEARDED DRAGON CARE

Feeding, Habitats, and Fun Tips for Reptile Lovers

Lily Dragonstone

Bearded Dragon Care

Feeding, Habitats, and Fun Tips for Reptile Lovers

Lily Dragonstone

Copyright

Copyright © 2024 by Lilly Dragonstone. All rights reserved.

No part of this book may be reproduced, distributed, or transmitted in any form or by any means, including photocopying, recording, or other electronic or mechanical methods, without the prior written permission of the author, except in the case of brief quotations embodied in critical reviews and certain other noncommercial uses permitted by copyright law.

Disclaimer

The information provided in this book is for educational purposes only and is not a substitute for professional veterinary advice. The author has made every effort to ensure accuracy but assumes no responsibility for errors or omissions. Always consult a qualified veterinarian for any health concerns regarding your bearded dragon. The author is not liable for any damages resulting from the use of the information in this book.

Introduction	7
Chapter 1: Getting to Know Your Bearded Dragon	**11**
Bearded Dragons	11
History and Origin	11
Types of Bearded Dragons	12
Understanding Their Behavior	13
Building a Relationship	14
Chapter 2: Preparing for Your Bearded Dragon	**17**
Preparation	17
Choosing the Right Bearded Dragon	17
Essential Supplies and Equipment	18
Setting Up the Perfect Habitat	19
Temperature and Lighting	20
Decorating the Enclosure	21
Final Checklist	22
Chapter 3: Setting Up the Perfect Habitat	**23**
Creating a Comfortable Home	23
Ideal Tank Size and Layout	25
Temperature and Lighting	26
How to Maintain the Correct Temperature Gradient	27
Decorating the Enclosure	27
Tips for Making the Habitat Aesthetically Pleasing and Functional	28

Chapter 4: Feeding Your Bearded Dragon	**31**
Feeding	31
Diet Basics	31
Feeding Juvenile vs. Adult Dragons	32
Insects and Vegetables	33
Supplements and Hydration	34
Feeding Tips and Tricks	35
Troubleshooting Feeding Issues	36
Chapter 5: Health and Wellness	**37**
Common Health Issues	37
Symptoms to Watch For and How to Prevent Them	38
Regular Health Checks	40
Finding a Reptile Vet	42
When to Seek Professional Help	43
Chapter 6: Handling and Bonding with Your Bearded Dragon	**46**
Safe Handling Techniques	46
Building Trust	48
Enrichment and Play	49
Chapter 7: Breeding Bearded Dragons	**52**
Introduction to Breeding	52
Breeding Setup	53
Chapter 8: Troubleshooting and FAQs	**59**
Common Problems and Solutions	59
Frequently Asked Questions	62
Conclusion	**66**
Summary of Key Points	66
Tips for Success	69

Introduction

Hello there, fellow reptile enthusiast!

I'm thrilled to welcome you to this journey into the wonderful world of bearded dragons. If you're anything like me, your fascination with these incredible creatures runs deep, sparking a joy that's hard to put into words. I remember the first time I laid eyes on a bearded dragon – it was love at first sight. The way they move, their quirky behaviors, and those expressive eyes that seem to understand so much more than we give them credit for – it's truly enchanting.

My own adventure with bearded dragons began a few years ago when a friend introduced me to her pet dragon named Spike. Spike was unlike any other pet I had encountered. He had a personality all his own – curious, friendly, and surprisingly affectionate. Watching him bask under his heat lamp or chase after crickets was mesmerizing. It wasn't long before I knew I had to have one of my own.

After welcoming my bearded dragon, Gizmo, into my home, I quickly realized there was so much to learn about caring for these unique pets. From setting up the perfect habitat to understanding their dietary needs, it felt like I was constantly uncovering new layers of information. It was both exciting and overwhelming. This book is a culmination of those experiences, the countless hours of research, and the deep bond I've formed with Gizmo.

I wanted to create a guide that feels like a conversation with a friend – someone who's been there, made a few mistakes, and come out the other side with a wealth of knowledge and a heart full of love for these amazing reptiles. Whether you're a seasoned dragon owner or just starting out, my hope is that you'll find this book both informative and comforting, a companion on your journey with your own bearded dragon.

Each chapter is crafted with care, filled with tips, stories, and insights that I've gathered along the way.

You'll find everything you need to know about setting up a welcoming home for your dragon, feeding them a nutritious diet, understanding their behaviors, and keeping them healthy and happy. And of course, there are plenty of fun tips for bonding with your new scaly friend.

As we delve into the world of bearded dragon care, I encourage you to keep an open heart and a curious mind. These creatures have so much to teach us about patience, responsibility, and the joy of connection. I'm excited to share this journey with you and can't wait to hear about the adventures you'll have with your own bearded dragon.

Let's get started on this exciting path together. Here's to the love of bearded dragons and the wonderful experiences ahead!

Warmest regards

Lily Dragonstone

Chapter 1: Getting to Know Your Bearded Dragon

Bearded Dragons

Welcome to the first chapter, where we'll dive into the fascinating world of bearded dragons. Before we get into the nitty-gritty of caring for these incredible reptiles, it's important to understand their origins, types, and behaviors. This foundational knowledge will help you appreciate your bearded dragon even more and ensure you're prepared to meet their needs.

History and Origin

Bearded dragons, or "Pogona," hail from the arid, rocky deserts of Australia. Their natural habitat is quite harsh, with extreme temperatures and scarce water sources. These resilient creatures have adapted beautifully to their environment, developing unique behaviors and physical traits that make them well-suited to desert life.

I remember reading about their natural habitat and being amazed at how different it is from my cozy home. It's incredible to think that these dragons, who now bask comfortably under heat lamps in our living rooms, once roamed the vast, wild landscapes of Australia.

Types of Bearded Dragons

There are several species of bearded dragons, each with its own unique characteristics. The most common pet species is the Central Bearded Dragon (Pogona vitticeps), known for its friendly demeanor and ease of care. Here are a few other notable types:

- Eastern Bearded Dragon (Pogona barbata): Known for their larger size and darker coloration.

- Rankin's Dragon (Pogona henrylawsoni): Smaller and more timid, often preferred by those with limited space.

- Western Bearded Dragon (Pogona minima): Less common in the pet trade, smaller and more slender.

Understanding the different species can help you appreciate the diversity within the bearded dragon family. Each type has its own charm, and knowing about them can help you make an informed decision if you're still considering which bearded dragon to bring home.

Understanding Their Behavior

One of the most delightful aspects of owning a bearded dragon is getting to know their unique behaviors. These reptiles are surprisingly expressive and can communicate a lot through their actions. Here are a few common behaviors you might observe:

- Head Bobbing: Often seen in males, this is a sign of dominance or mating behavior.

- Arm Waving: A submissive gesture, often seen in females or younger dragons.

- Basking: Bearded dragons love to bask under their heat lamp, soaking up the warmth.

- Glass Surfing: When they scratch or "surf" against the glass of their tank, usually a sign of stress or boredom.

I'll never forget the first time Gizmo started waving his little arm at me. At first, I was worried something was wrong, but after a bit of research, I realized he was just being polite in his dragon way! Observing these behaviors can help you better understand your bearded dragon's needs and emotions.

Building a Relationship

Developing a bond with your bearded dragon takes time and patience, but it's incredibly rewarding. Spend time each day interacting with your dragon, whether it's through gentle handling, feeding, or simply talking to them. They might not understand the words, but they'll recognize your voice and presence.

As you continue to care for your bearded dragon, you'll find that they have their own unique personality. Some are more outgoing and curious, while others might be shy and reserved. Respect their individuality and adjust your interactions accordingly.

Getting to know your bearded dragon is the first step in building a strong, loving relationship. With this foundational understanding, you're now ready to prepare for their arrival. In the next chapter, we'll explore everything you need to know about setting up a welcoming and comfortable habitat for your new scaly friend.

Chapter 2: Preparing for Your Bearded Dragon

Preparation

Preparing for a new bearded dragon is an exciting and crucial step in ensuring their health and happiness. This chapter will guide you through choosing the right dragon, gathering essential supplies, and setting up a habitat that mimics their natural environment as closely as possible. Let's get started on making your home a welcoming haven for your new scaly friend.

Choosing the Right Bearded Dragon

When I was choosing Gizmo, my bearded dragon, I spent countless hours researching and visiting different breeders and pet stores. The experience was both thrilling and a bit overwhelming. Here are some tips to help you choose the right bearded dragon:

- Health Check: Look for a dragon with clear, bright eyes, a full tail, and no visible signs of injury or illness. Healthy dragons are alert and active.

- Age and Size: Decide whether you want a juvenile or an adult. Juveniles require more attention and careful feeding, while adults are generally more robust.

- Temperament: Spend some time observing the dragon's behavior. A calm, curious dragon is usually a good sign.

Essential Supplies and Equipment

Before bringing your bearded dragon home, you'll need to gather some essential supplies. Here's a list to get you started:

- Enclosure: A 40-gallon tank is a good starting size for a juvenile, but adults need at least a 75-gallon tank.

- Lighting: UVB lighting is crucial for their health. A high-quality UVB bulb and a basking light are necessary.

- Heating: Bearded dragons need a temperature gradient in their tank. Invest in a reliable heat lamp and a thermostat to regulate temperatures.

- Substrate: Choose a safe substrate like reptile carpet or tile. Avoid loose substrates like sand, which can cause impaction.

- Hides and Decorations: Provide hiding spots and climbing structures to keep your dragon stimulated.

- Food and Water Dishes: Shallow dishes for water and food that are easy to clean.

Setting Up the Perfect Habitat

Creating a comfortable and safe habitat is essential for your bearded dragon's well-being. Here's a step-by-step guide to setting up their new home:

1. Clean the Enclosure: Before adding any items, thoroughly clean the tank with a reptile-safe disinfectant.

2. Install Lighting and Heating: Place the UVB light and basking lamp on one side of the tank to create a temperature gradient. Ensure the basking spot reaches around 95-105°F, while the cooler side should be about 75-85°F.

3. Add Substrate: Lay down the substrate, making sure it covers the bottom of the tank evenly.

4. Decorate the Enclosure: Add hides, branches, and rocks to create a stimulating environment. Ensure that all decorations are stable and won't fall over.

5. Set Up Food and Water Dishes: Place shallow dishes for water and food in the tank. Clean and refill them daily.

Temperature and Lighting

Maintaining proper temperature and lighting is crucial for your bearded dragon's health. They rely on external heat sources to regulate their body temperature and digest food properly.

- Basking Spot: Ensure the basking spot is between 95-105°F. Use a thermometer to monitor the temperature.

- Cool Side: The cool side of the tank should be around 75-85°F. This gradient allows your dragon to thermoregulate.

- UVB Lighting: UVB rays are essential for calcium metabolism and preventing metabolic bone disease. Replace the UVB bulb every six months, even if it still produces light.

Decorating the Enclosure

Creating an enriching environment for your bearded dragon is both fun and important for their mental stimulation. Here are some decoration ideas:

- Hides: Provide at least two hides – one on the warm side and one on the cool side of the tank.

- Branches and Rocks: These offer climbing opportunities and help with shedding.

- Plants: Use reptile-safe artificial plants to add some greenery.

Final Checklist

Before bringing your bearded dragon home, go through this final checklist to ensure everything is ready:

- Enclosure is clean and set up
- Lighting and heating are functioning correctly
- Substrate is laid down
- Decorations and hides are secure
- Food and water dishes are in place

Preparing for your bearded dragon's arrival is an exciting time filled with anticipation. By ensuring their habitat is set up correctly, you're laying the foundation for a happy and healthy life for your new pet. In the next chapter, we'll delve into the specifics of feeding your bearded dragon, ensuring they receive a balanced and nutritious diet.

Chapter 3: Setting Up the Perfect Habitat

Creating the perfect habitat for your bearded dragon is essential to their health, happiness, and overall well-being. A well-designed enclosure mimics their natural environment, providing the necessary conditions for them to thrive. In this chapter, we'll guide you through setting up a comfortable and stimulating home for your new scaly friend.

Creating a Comfortable Home

Step-by-Step Guide to Setting Up the Tank

1. Choose the Right Tank: For a juvenile bearded dragon, a 40-gallon tank is a good starting point. However, as they grow, they'll need more space. An adult bearded dragon requires at least a 75-gallon tank.

Glass terrariums are a popular choice because they provide good visibility and are easy to clean.

2. Clean the Tank: Before setting up, thoroughly clean the tank with a reptile-safe disinfectant. This ensures that any harmful bacteria or chemicals are removed.

3. Install the Substrate: The substrate is the material that lines the bottom of the tank. Choose a safe substrate such as reptile carpet, paper towels, or non-adhesive shelf liner. Avoid loose substrates like sand or wood chips, which can cause impaction if ingested.

4. Set Up Heating and Lighting: Place a basking lamp on one side of the tank to create a warm basking area, and install a UVB light that spans at least two-thirds of the enclosure. Both are crucial for your dragon's health and well-being.

5. Arrange Hides and Climbing Structures: Provide at least two hides—one on the warm side and one on the cool side of the tank. Include branches, rocks, and other climbing structures to mimic their natural habitat and offer mental stimulation.

6. Place Food and Water Dishes: Use shallow dishes for food and water, ensuring they are easy to clean and maintain.

7. Monitor and Adjust: Regularly check the tank's temperature and humidity levels, making adjustments as needed to maintain optimal conditions.

Ideal Tank Size and Layout

- Tank Size: As mentioned, a juvenile needs at least a 40-gallon tank, while an adult requires a minimum of 75 gallons. Larger is always better if space allows.

- Layout: Design the layout to include distinct basking and cool areas. Place the basking lamp on one end of the tank and provide hides on both ends to allow your dragon to regulate its body temperature.

Temperature and Lighting

Importance of Proper Heating and UVB Lighting

Bearded dragons are ectothermic, meaning they rely on external heat sources to regulate their body temperature.
Proper heating and UVB lighting are essential for their digestion, immune system, and overall health.

- Basking Spot: The basking spot should reach 95-105°F during the day. This area allows your dragon to warm up and digest food properly.

- Cool Side: The cool side of the tank should be around 75-85°F. This gradient enables your dragon to move between warm and cool areas as needed.

- Nighttime Temperature: At night, temperatures can drop to 65-75°F. Use a ceramic heat emitter if additional heat is needed without producing light.

How to Maintain the Correct Temperature Gradient

- Thermometers: Place thermometers on both the basking and cool sides of the tank to monitor temperatures accurately.

- Adjustable Lamps: Use lamps with adjustable stands to raise or lower them, controlling the temperature.

- Timers: Set up timers for the lights to ensure your dragon has a consistent day/night cycle, typically 12 hours of light and 12 hours of darkness.

Decorating the Enclosure

Safe and Stimulating Decorations

Decorating your bearded dragon's enclosure is not just about aesthetics—it's also about providing a stimulating environment that encourages natural behaviors.

- Hides: Provide at least two hides, one on the warm side and one on the cool side of the tank.

These give your dragon a sense of security and a place to retreat.

- Climbing Structures: Include branches, rocks, and other items that encourage climbing. These help with physical exercise and mental stimulation.

- Plants: Use artificial plants to add greenery. Ensure they are securely attached and can't be ingested.

Tips for Making the Habitat Aesthetically Pleasing and Functional

- Natural Look: Aim for a naturalistic setup that mimics the bearded dragon's desert environment. Use materials like rocks, driftwood, and sand-colored substrates.

- Easy Cleaning: Arrange decorations in a way that allows for easy access when cleaning the tank. Removable items make it easier to maintain hygiene.

- Personal Touch: Add personal touches that reflect your style, while ensuring all items are safe and non-toxic for your dragon.

Creating a comfortable and stimulating habitat is the foundation for a healthy and happy bearded dragon. By carefully setting up their environment, you'll provide a space where they can thrive and display natural behaviors. In the next chapter, we'll delve into the specifics of feeding your bearded dragon, ensuring they receive a balanced and nutritious diet.

Chapter 4: Feeding Your Bearded Dragon

Feeding

Feeding your bearded dragon is one of the most important aspects of their care. A balanced diet ensures they stay healthy, active, and happy. In this chapter, we'll explore what to feed your bearded dragon, how often to feed them, and tips for maintaining a nutritious diet that mimics what they'd find in the wild.

Diet Basics

Understanding the basic dietary needs of your bearded dragon is the first step in providing proper care. Bearded dragons are omnivores, meaning they eat both animal and plant matter. Their diet typically consists of:

- Insects: Crickets, mealworms, and dubia roaches are popular choices.

- Vegetables: Leafy greens like collard greens, mustard greens, and dandelion greens.

- Fruits: Occasional treats like blueberries, strawberries, and apples.

When I first brought Gizmo home, I was overwhelmed by the variety of food options. It took some trial and error to figure out what he liked and what worked best for his health. With a little patience, you'll quickly become an expert on your dragon's preferences.

Feeding Juvenile vs. Adult Dragons

The dietary needs of bearded dragons change as they grow. Juvenile dragons (up to 12 months old) require more protein to support their rapid growth, while adults need a balanced diet to maintain their health.

- Juveniles: 70% insects, 30% vegetables
- Adults: 30% insects, 70% vegetables

Juveniles should be fed insects 2-3 times a day, with a constant supply of fresh vegetables.

Adults can be fed insects once a day or every other day, with daily servings of vegetables.

Insects and Vegetables

Choosing the right insects and vegetables is crucial for your dragon's health. Here are some tips:

- Insects: Offer a variety of insects to ensure a balanced diet. Dust insects with calcium powder to prevent metabolic bone disease. Avoid wild-caught insects, as they may carry parasites.

- Vegetables: Rotate different types of greens to provide a range of nutrients. Avoid spinach and kale in large amounts, as they can bind calcium.

Safe Insects:
- Crickets
- Dubia roaches
- Mealworms (in moderation)

Safe Vegetables:
- Collard greens
- Mustard greens

- Dandelion greens
- Squash
- Bell peppers

Supplements and Hydration

Supplements are essential for your bearded dragon's health, particularly calcium and vitamins. Dust their food with calcium powder 3-4 times a week and a multivitamin once a week.

Hydration is also important. While bearded dragons get most of their water from food, they still need a water dish. Mist their vegetables lightly to ensure they're getting enough moisture. Some dragons enjoy soaking in a shallow water dish, which also helps with shedding.

Feeding Tips and Tricks

Feeding your bearded dragon can be a fun and interactive experience. Here are some tips to make it easier:

- Routine: Establish a regular feeding schedule to keep your dragon healthy and happy.

- Variety: Mix up the types of insects and vegetables to keep meals interesting.

- Hand Feeding: Occasionally hand feed your dragon to build trust and strengthen your bond.

- Observation: Pay attention to their eating habits and preferences. If they're not eating well, it could be a sign of illness.

One of my favorite memories with Gizmo is discovering his love for blueberries. Watching him chase after them and munch happily was a joy. These moments of connection over food are truly special and help build a strong bond with your dragon.

Troubleshooting Feeding Issues

Sometimes, bearded dragons can be picky eaters or may refuse food. Here are some common issues and solutions:

- Stress: New environments or changes in routine can cause stress. Give them time to adjust.

- Temperature: Ensure their habitat is at the correct temperature. If it's too cold, they may not eat.

- Health Issues: If they consistently refuse food, consult a vet to rule out any underlying health problems.

Feeding your bearded dragon a balanced and nutritious diet is key to their overall health and happiness. By providing a variety of foods and paying attention to their preferences, you can ensure they thrive. In the next chapter, we'll explore how to keep your bearded dragon healthy and what to do if they fall ill.

Chapter 5: Health and Wellness

Ensuring the health and wellness of your bearded dragon is a crucial aspect of pet ownership. By understanding common health issues, performing regular health checks, and knowing when to seek professional help, you can provide the best care for your scaly friend. This chapter will guide you through the essentials of keeping your bearded dragon healthy and happy.

Common Health Issues

Bearded dragons, like all pets, can experience health problems. Being aware of the symptoms and knowing how to prevent common issues can make a significant difference in their quality of life.

Symptoms to Watch For and How to Prevent Them

1. Metabolic Bone Disease (MBD):

   - Symptoms: Weakness, tremors, soft jaw or limbs, and difficulty moving.

   - Prevention: Ensure your dragon receives adequate UVB lighting and calcium supplements. Provide a balanced diet rich in calcium.

2. Respiratory Infections:

   - Symptoms: Wheezing, mucus around the nostrils, open-mouth breathing, and lethargy.

   - Prevention: Maintain proper tank temperatures and humidity levels. Avoid exposing your dragon to cold drafts.

3. Impaction:

   - Symptoms: Lack of bowel movements, bloating, and decreased appetite.

- Prevention: Avoid using loose substrates like sand. Ensure they consume appropriate food sizes to prevent blockages.

4. Parasites:

- Symptoms: Weight loss, abnormal stools, and lethargy.

- Prevention: Regularly clean the tank and provide a balanced diet. Schedule routine vet check-ups to catch any issues early.

5. Dehydration:

- Symptoms: Sunken eyes, wrinkled skin, and lack of appetite.

- Prevention: Provide fresh water daily, mist vegetables, and occasionally offer baths.

One of the scariest moments I had with Gizmo was when he showed signs of a respiratory infection. I noticed he was breathing with his mouth open and seemed unusually sluggish. Thanks to quick action and a visit to a knowledgeable reptile vet, Gizmo recovered fully. These experiences underline the importance of vigilance and prompt care.

Regular Health Checks

Performing routine health assessments at home can help you catch potential issues early. Here's how to do a basic health check:

How to Perform Routine Health Assessments at Home

1. Observe Behavior: Watch for any changes in activity levels, appetite, or behavior. Lethargy or hiding more than usual can be signs of illness.

2. Examine Eyes: Ensure your dragon's eyes are bright and clear. Cloudy or sunken eyes can indicate health issues.

3. Check Skin and Scales: Look for signs of shedding problems, cuts, or abnormal growths. Healthy skin should be smooth and free of lesions.

4. Monitor Weight: Regularly weigh your dragon to ensure they're maintaining a healthy weight. Sudden weight loss can be a red flag.

5. Inspect the Mouth: Gently open their mouth to check for discoloration or sores. A healthy mouth is pink without any signs of infection.

6. Check Limbs and Tail: Ensure there are no signs of swelling, deformities, or injuries. Limping or dragging limbs can indicate problems.

Performing these checks weekly will help you stay on top of your bearded dragon's health. Keep a journal to note any changes or concerns, which can be useful during vet visits.

Finding a Reptile Vet

Not all veterinarians are experienced with reptiles, so finding a qualified reptile vet is essential for your bearded dragon's health. Here's how to choose the right vet and when to seek professional help.

Tips for Choosing a Qualified Veterinarian

1. Research: Look for vets in your area who specialize in reptiles or have experience treating them.

2. Ask for Recommendations: Reach out to local reptile clubs, pet stores, or other dragon owners for vet recommendations.

3. Check Credentials: Ensure the vet is licensed and has specific training or certifications in reptile care.

4. Visit the Clinic: Schedule a visit to see the clinic's facilities and ask the vet about their experience with bearded dragons.

5. Trust Your Instincts: Choose a vet who communicates well, answers your questions thoroughly, and makes you feel comfortable.

When to Seek Professional Help

- Persistent Symptoms: If your dragon shows signs of illness that don't improve with home care.

- Injuries: Any injuries that cause bleeding, swelling, or pain.

- Behavioral Changes: Sudden changes in behavior, such as refusal to eat or move.

- Routine Check-Ups: Schedule regular check-ups to monitor your dragon's health and catch any issues early.

Taking care of your bearded dragon's health and wellness requires dedication, observation, and a willingness to seek help when needed. By staying informed about common health issues, performing regular checks, and finding a reliable reptile vet, you'll ensure your dragon lives a long, happy, and healthy life. In the next chapter, we'll explore how to handle and bond with your bearded dragon, creating a strong and loving relationship.

Chapter 6: Handling and Bonding with Your Bearded Dragon

Creating a strong bond with your bearded dragon is one of the most rewarding aspects of pet ownership. While they may not be as overly affectionate as dogs or cats, bearded dragons have their own unique ways of showing trust and attachment. In this chapter, we'll explore safe handling techniques, tips for building trust, and fun activities to keep your dragon engaged and happy.

Safe Handling Techniques

Proper handling is essential for both your safety and the well-being of your bearded dragon. Here's how to handle your dragon safely and confidently:

1. Approach Calmly: Always approach your bearded dragon slowly and from the side. Sudden movements can startle them.

2. Support Their Body: Use both hands to lift your dragon, supporting their chest and hindquarters. Never grab them by the tail or limbs, as this can cause injury.

3. Close to the Ground: When handling your dragon for the first few times, stay close to the ground to prevent injury if they jump or fall.

4. Gently and Firmly: Hold them gently but securely. Avoid squeezing too hard, but ensure they can't wriggle free and fall.

5. Watch for Stress Signs: If your dragon puffs up, hisses, or tries to escape, it may be stressed. Return them to their enclosure and try again later.

Handling Gizmo for the first time was a mix of excitement and nervousness. I remember being so careful, afraid I might hurt him. With time and practice, handling him became second nature, and I could see how much he enjoyed our interactions.

Building Trust

Trust is the foundation of a strong bond with your bearded dragon. Here are some tips to help you build trust and create a positive relationship:

1. Consistent Interaction: Spend time with your dragon every day. Regular interaction helps them get used to your presence and voice.

2. Hand Feeding: Hand feeding treats is a great way to build trust. It encourages your dragon to associate your hand with positive experiences.

3. Gentle Talking: Talk to your dragon in a calm, soothing voice. They may not understand the words, but they'll recognize and become comforted by your voice.

4. Patience: Building trust takes time. Be patient and respect your dragon's boundaries. Forcing interactions can lead to stress and fear.

5. Positive Reinforcement: Reward good behavior with treats and gentle petting. Positive reinforcement helps your dragon associate handling with positive outcomes.

One of my fondest memories is hand feeding Gizmo his favorite treats. Watching his little tongue dart out and gently take the food from my hand always brought a smile to my face. These moments helped solidify our bond and made handling him much easier.

Enrichment and Play

Keeping your bearded dragon mentally and physically stimulated is crucial for their well-being. Here are some fun activities and enrichment ideas to keep your dragon engaged:

1. Exploration Time: Allow your dragon supervised time outside of their enclosure. Create a safe space where they can explore and roam.

2. Climbing Structures: Provide branches, rocks, and other climbing structures inside the enclosure. These encourage natural climbing behaviors.

3. Interactive Toys: Offer safe, interactive toys like plastic balls or reptile-safe objects they can push around.

4. Obstacle Courses: Set up mini obstacle courses with tunnels and ramps. This provides physical exercise and mental stimulation.

5. Bath Time: Many bearded dragons enjoy soaking in shallow water. This helps with hydration and shedding, and can be a fun activity.

I remember setting up a small play area for Gizmo with tunnels and ramps. Watching him explore and navigate the obstacles was not only entertaining for him but also for me. These activities help keep your dragon active and engaged, reducing boredom and promoting overall health.

Handling and bonding with your bearded dragon is a journey filled with rewarding experiences and deep connections.

By following safe handling techniques, building trust through consistent interaction, and providing enriching activities, you'll create a strong and loving relationship with your scaly companion. In the next chapter, we'll explore the basics of breeding bearded dragons, from preparation to caring for eggs and hatchlings.

Chapter 7: Breeding Bearded Dragons

Breeding bearded dragons can be a fascinating and rewarding endeavor, but it requires careful preparation, knowledge, and commitment. This chapter will guide you through the basics of breeding bearded dragons, including preparation, breeding setup, and caring for eggs and hatchlings.

Introduction to Breeding

Breeding bearded dragons is not for everyone, and it's important to understand the responsibility it entails. Before deciding to breed your dragons, consider the following:

- Time Commitment: Breeding and caring for hatchlings requires a significant amount of time and effort.

- Space: Ensure you have enough space to house additional dragons and hatchlings.

- Knowledge: Educate yourself thoroughly about the breeding process, potential complications, and proper care.

If you're ready to proceed, breeding can be an incredibly rewarding experience, providing insight into the life cycle of these amazing creatures.

Breeding Setup

Creating the right environment is crucial for successful breeding. Here's a step-by-step guide to setting up for breeding:

1. Choosing Breeding Pairs:

   - Health and Age: Ensure both dragons are healthy and at least 18 months old. Breeding younger dragons can cause health issues.

   - Compatibility: Introduce the male and female in a neutral space to observe their behavior. Signs of aggression indicate they may not be compatible.

2. Setting Up the Breeding Enclosure:

- Size: Use a large enclosure (at least 75 gallons) to provide ample space.

- Substrate: Use a mixture of soil and sand to create a digging area for the female to lay eggs.

- Temperature and Lighting: Maintain proper temperatures with a basking spot (95-105°F) and a cool side (75-85°F). Ensure adequate UVB lighting.

3. Courtship and Mating:

- Introduction: Place the male in the female's enclosure for short periods to allow for courtship behaviors. This may include head bobbing, arm waving, and chasing.

- Mating: Once the female is receptive, the male will mount her and bite the back of her neck to hold her in place. This is normal behavior and usually lasts a few minutes.

Caring for Eggs and Hatchlings

After successful mating, the female will lay eggs within a few weeks. Here's how to care for the eggs and hatchlings:

1. Preparing the Lay Box:

   - Size and Material: Use a large plastic container filled with a mixture of soil and sand.

   - Placement: Place the lay box in the female's enclosure and ensure it's accessible.

2. Egg Laying:

   - Signs: The female will become restless and start digging. She may eat less and spend more time in the lay box.

   - Laying: Once she begins laying, leave her undisturbed. The process can take several hours.

3. Incubating the Eggs:

   - Collection: Gently remove the eggs and place them in an incubation container filled with moist vermiculite.

- Incubator Setup: Maintain a temperature of 82-86°F and humidity around 75-80%. Use an incubator to ensure consistent conditions.

- Monitoring: Check the eggs regularly for mold or collapse. Healthy eggs will remain plump and firm.

4. Hatching:

- Time Frame: Eggs typically hatch after 55-75 days.

- Signs of Hatching: Hatchlings will use their egg tooth to break through the shell. This process can take up to 24 hours.

- First Days: Leave the hatchlings in the incubator for a day or two before transferring them to a rearing enclosure.

5. Caring for Hatchlings:

- Rearing Enclosure: Use a small, separate enclosure for hatchlings. Maintain proper temperatures, lighting, and humidity.

- Diet: Offer small, appropriately sized insects and finely chopped vegetables. Dust insects with calcium powder.

- Hydration: Provide shallow water dishes and mist the enclosure to ensure proper hydration.

Breeding bearded dragons is a complex but rewarding process that requires dedication and knowledge. By preparing adequately, creating the right environment, and providing proper care for eggs and hatchlings, you can ensure a successful breeding experience. Remember, breeding should always be undertaken responsibly, with the well-being of the dragons as the top priority.

In the next chapter, we'll address common problems and frequently asked questions to help you troubleshoot any issues and provide the best care for your bearded dragon.

Chapter 8: Troubleshooting and FAQs

Even with the best preparation and care, you may encounter challenges while caring for your bearded dragon. This chapter addresses common problems and provides solutions, as well as answers to frequently asked questions to help you navigate any issues and ensure your dragon remains healthy and happy.

Common Problems and Solutions

1. Refusal to Eat:

   - Possible Causes: Stress, improper temperatures, illness, or changes in environment.

   - Solutions: Check and adjust habitat conditions. Ensure the enclosure is at the correct temperature. Try offering different types of food. If the problem persists, consult a vet.

2. Shedding Issues:

   - Possible Causes: Low humidity, poor diet, or lack of hydration.

   - Solutions: Increase humidity by misting the enclosure or providing a humid hide. Offer a bath to help loosen the skin. Ensure a balanced diet with proper vitamins and minerals.

3. Behavioral Changes:

   - Possible Causes: Stress, illness, changes in environment, or breeding season.

   - Solutions: Identify and minimize stressors. Ensure the habitat is properly set up. If the dragon is showing signs of illness, consult a vet. During the breeding season, some changes in behavior are normal.

4. Respiratory Issues:

   - Possible Causes: Improper temperatures, high humidity, or infections.

   - Solutions: Ensure proper temperature gradient and ventilation. Keep humidity levels appropriate. If symptoms persist, seek veterinary care.

5. Impaction:

   - Possible Causes: Ingesting substrate, eating large food items, or dehydration.

   - Solutions: Avoid loose substrates. Ensure food is appropriately sized. Provide regular hydration through misting and baths. If signs of impaction occur (lethargy, lack of bowel movements), consult a vet immediately.

One time, Gizmo refused to eat for several days. I was worried sick, but after adjusting his tank's temperature and offering his favorite food, he perked up and started eating again. It was a relief and a reminder of how small changes can significantly impact their well-being.

Frequently Asked Questions

Q1: How often should I feed my bearded dragon?

- A: Juveniles should be fed insects 2-3 times a day, while adults can be fed insects once a day or every other day. Vegetables should be offered daily for both juveniles and adults.

Q2: How do I know if my bearded dragon is healthy?

- A: A healthy bearded dragon is active, alert, and has clear eyes and smooth skin. They should have a good appetite and display normal behaviors like basking and exploring.

Q3: What are the best substrates for my bearded dragon?

- A: Safe substrates include reptile carpet, paper towels, and non-adhesive shelf liner. Avoid loose substrates like sand, which can cause impaction.

Q4: How can I tell if my bearded dragon is too hot or too cold?

- A: Monitor the temperatures in their enclosure. Signs of overheating include open-mouth breathing and excessive time spent in the cool area. Signs of being too cold include lethargy and lack of appetite.

Q5: Can I keep more than one bearded dragon together?

- A: It's generally not recommended to house multiple bearded dragons together, as they can become territorial and aggressive. If you do, ensure a large enclosure and monitor them closely.

Q6: What should I do if my bearded dragon has parasites?

- A: If you suspect parasites (e.g., weight loss, abnormal stools), consult a reptile vet for a proper diagnosis and treatment plan. Keep the enclosure clean to prevent infestations.

Q7: How can I tell if my bearded dragon is stressed?

- A: Signs of stress include glass surfing, hiding excessively, changes in eating habits, and dark stress marks on their belly. Identify and reduce stressors, such as changes in environment or handling too frequently.

Q8: How often should I clean my bearded dragon's enclosure?

- A: Spot-clean daily by removing waste and uneaten food. Perform a thorough cleaning, including disinfecting the tank and decorations, every 4-6 weeks.

Troubleshooting common problems and understanding frequently asked questions are essential for providing excellent care for your bearded dragon. With the knowledge and solutions provided in this chapter, you'll be better equipped to handle any challenges that arise and ensure your dragon remains healthy and happy.

Conclusion

Congratulations on reaching the end of this comprehensive guide to bearded dragon care! By now, you should have a solid understanding of what it takes to keep your bearded dragon healthy, happy, and thriving. Let's recap the key points and leave you with some final tips for success.

Summary of Key Points

1. Understanding Your Bearded Dragon:

   - Bearded dragons are fascinating reptiles from the arid regions of Australia.

   - There are several species, each with unique characteristics.

2. Preparing for Your Bearded Dragon:

   - Choose a healthy, well-adjusted bearded dragon.

   - Gather essential supplies, including a properly sized tank, lighting, heating, and substrate.

3. Setting Up the Perfect Habitat:

   - Create a comfortable home with the right temperature gradient and UVB lighting.

   - Decorate the enclosure with safe and stimulating items like hides, branches, and plants.

4. Feeding Your Bearded Dragon:

   - Provide a balanced diet of insects, vegetables, and occasional fruits.

   - Adjust the diet according to the dragon's age and nutritional needs.

5. Health and Wellness:

   - Monitor for common health issues and perform regular health checks.

   - Find a qualified reptile vet and know when to seek professional help.

6. Handling and Bonding:

   - Use safe handling techniques and build trust through consistent interaction.

   - Engage your dragon with enrichment activities and play.

7. Breeding Bearded Dragons:

   - Prepare adequately for breeding, create the right environment, and care for eggs and hatchlings responsibly.

8. Troubleshooting and FAQs:

   - Address common problems with practical solutions.

   - Refer to the FAQs for quick answers to typical questions.

Tips for Success

1. Stay Informed:
   - Continuously educate yourself about bearded dragons. Join online forums, read books, and connect with other owners.

2. Patience and Consistency:
   - Building a bond with your bearded dragon takes time. Be patient and consistent in your care and interactions.

3. Regular Vet Visits:
   - Schedule regular check-ups with a reptile vet to ensure your dragon's health.

4. Observe and Adapt:
   - Pay close attention to your dragon's behavior and make adjustments to their care as needed. Each dragon is unique and may have specific preferences and needs.

5. Enjoy the Journey:
   - Caring for a bearded dragon is a rewarding experience. Enjoy the journey and cherish the moments you share with your scaly friend.

I hope this guide has provided you with valuable insights and practical advice to help you care for your bearded dragon. Whether you're a first-time owner or an experienced enthusiast, my wish is that you feel confident and prepared to provide the best possible care for your pet.

Thank you for choosing to embark on this journey with me. I wish you and your bearded dragon many happy, healthy years together!

Warmest regards,

Lily Dragonstone

www.ingramcontent.com/pod-product-compliance
Lightning Source LLC
Chambersburg PA
CBHW071957210526
45479CB00003B/967